GARFIELD COUNTY LIBRARIES
Carbondale Branch Library
320 Sopris Ave
Carbondale, CO 81623
(970) 963-2889 – Fax (970) 963-8573
www.gcpld.org

Decorative Papers
made easy

Decorative Papers
made easy

Series Editors: Susan & Martin Penny

David & Charles

A DAVID & CHARLES BOOK

First published in the UK in 2000

A catalogue record for this book is available from the British Library.

ISBN 0 7153 1016 X

Series Editors: Susan & Martin Penny
Designed and produced by Penny & Penny
Illustrations: Fred Fieber at Red Crayola
Photography: John Gollop

Printed in France by Imprimerie Pollina S. A.
for David & Charles
Brunel House Newton Abbot Devon

Contents

Introduction to Decorative Papers

Decorative Papers Made Easy is a complete guide to the craft of making and decorating your own paper; find out how to make textured, decorated and coloured paper using scrap paper and ingredients from your kitchen. The paper can then be used in lots of imaginative ways from covering books and boxes to making lampshades and gift bags

Essential equipment

Below is a list of equipment needed for paper making and decorating:

- **Mould and Deckle** – two identical wooden frames, one covered with net, that fit exactly together and are used to collect paper pulp from the vat.
- **Vat** – a plastic tray larger than the mould and deckle, that will hold sufficient paper pulp to make a sheet of paper.
- **Bucket** – used for soaking scrap paper to make paper pulp.
- **Length of wood or hand liquidizer** – used for beating the paper pulp to a smooth, creamy paste.
- **Hardboard** – used to sandwich the newly made stack of paper.
- **Clean bricks or heavy weights** – used on top of the hardboard to apply pressure to the stack of paper.
- **Kitchen cloths** – used between each sheet of paper to soak up the water, and make it easier to move while wet.
- **Plastic sheet** – to cover your work surface.
- **Newspaper** – to protect your work surface.
- **Rolling pin** – to give the wet sheets of paper a smooth finish.
- **Paintbrush** – for applying paint over batik wax and to decorate paste paper.
- **Saucepan or Double Boiler** – for making vegetable paper, and heating batik wax.
- **Empty pet food tin** – for melting and pouring batik wax.

Making coloured paper

Paper pulp can be made from coloured waste paper; or from plain pulp coloured with a small amount of strong coloured paper, natural colouring or dye.

- **Recycling coloured paper**

  Use crackers, napkins, tissue, crepe, wrapping paper and paper bags.

- **Colouring the pulp**

  Use juice from soft fruit like blackberries, raspberries, strawberries and blackcurrants. Vegetable and carrot juice, boiled onion skins, beetroot juice.

  Spices like curry powder, turmeric, chilli powder and cinnamon.

  Leaf and fruit tea.

  Instant coffee.

  Powdered poster paint and drawing ink.

  Food colouring both powdered and liquid.
- **Colouring the finished paper**

  Food colouring.

  Fruit and vegetable juice.

  Poster paint, silk paint, acrylic paint and ink.

Tips for making paper

- ✔ Use a plastic sheet and plenty of newspaper to protect your work surface
- ✔ Clean the mould and deckle thoroughly between making each sheet of paper
- ✔ Use a cat litter tray as a vat for making paper and for marbling
- ✔ Let as much water drain away as possible before you transfer the newly made sheet of paper to the kitchen cloth
- ✔ Adjust the ratio of pulp to water in the vat between making each sheet, so that you have one third pulp to two thirds water.

Textured paper

Use some of these techniques to create paper with a textured surface.

- ✔ Leave the paper to dry on textured fabric like tweed, a lace mat or patterned net curtain
- ✔ Push objects into the damp paper to make a raised pattern
- ✔ Add dried lentils and split peas
- ✔ Add scraps of material, wool, raffia and ribbon to the pulp
- ✔ Add ribbon roses, beads, metal foil and scraps of paper to the surface of the pulp
- ✔ Drop pressed flowers and leaves, seeds and potpourri on to the pulp

Surface decoration

- ● **Marbling** – use marbling ink or oil paint to create a swirly pattern on the paper.
- ● **Batik** – use melted wax and paint to create a resist pattern.
- ● **Paste** – scratch a pattern on the paper in a layer of wallpaper paste and paint.
- ● **Fold 'n' dye** – Fold Japanese paper, then dip it in dye or paint.

Making pulp

How to choose the right paper

- ● Recycle paper with long fibres, as this will make the strongest paper.
- ● Tear the paper to see if there are wispy fibres on the tear: these are the long fibres.
- ● Avoid heavily printed paper, glossy magazines or mailshots.
- ● Newspaper should be avoided because the printing ink creates a black scum on the surface of the pulp.
- ● Avoid paper with a shiny surface as it may be coated with clay, and will leave powdery patches on the finished paper.
- ● Do not use paper that has been glued.
- ● Remove tape and staples before using the paper.

Types of paper to recycle

- ● Computer print-out paper
- ● Old envelopes
- ● Paper bags
- ● Wrapping paper
- ● Paper napkins
- ● Tissue paper
- ● Christmas decorations (crackers and paper hats, streamers)
- ● Crepe paper
- ● Photocopier paper

Making paper from plants

- ● Use plants with stringy, narrow leaves like rhubarb, celery, cow parsley, leeks, fennel, rushes, reeds, grasses, irises and daffodil leaves.

- ● Use pineapple tops, cabbage peelings, cauliflower leaves, onion skins and nettles.

Making a Mould and Deckle

To make paper you will need to buy or make a mould and deckle. These two identical wooden frames, one covered with net, act like a sieve collecting paper pulp whilst draining away the water. Use old curtain net or fine dressmaking net to cover the mould, stretching it tightly before stapling it at the edges

1 Cut four pieces of 2x1cm (³/₄x³/₈in) thick timber 27cm (10³/₄in) long, and four pieces 20cm (8in) long. Arrange two longer, and two shorter pieces together to make the mould. Use PVA glue to hold the pieces together. Use the other four pieces to make the deckle.

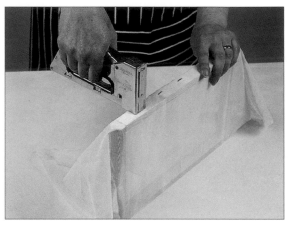

3 Cut a piece of fine net to fit over the mould. Wet the net and attach it to the frame using steel staples. Start by stapling the middle of each side, pulling as tightly as you can. Staple around the frame pulling the net taut. As the net dries it will become tighter.

2 Once the glue is dry, hammer two rust proof nails into each corner of the mould, making sure they are long enough to go into both lengths of wood. Repeat for the deckle.

4 Cut away the excess net, and then seal the edges with waterproof PVA glue. The deckle and mould should be exactly the same size when held together.

Making a Sheet of Paper

Almost any paper can be recycled, but pulp containing long fibres makes the best paper. Computer print-out paper, paper bags and envelopes are excellent – because they need to be strong they are made from long fibres. To check the length of the fibres, tear the paper and if it has long wispy ends it will make good paper

1 Prepare the paper to be recycled by removing all traces of glue and taking out any staples. Avoid paper with shiny surfaces as this can cause powdery patches on the finished paper. Tear the paper into small squares about the size of a postage stamp.

3 The wet paper needs to be beaten to a mushy pulp using a piece of wood or a hand held liquidizer. If there is too much water remaining in the bucket, tip some away before beating. The pulp needs to be very smooth and creamy.

2 Put the torn paper into a bucket of water and leave to soak for a few days. You may need to top up the water as it gets drawn into the paper.

4 Half fill a plastic tray with water, add the pulp and stir well until mixed. Adjust the ratio of pulp to water until you have one third pulp to two thirds water.

5 Paper making uses a lot of water, so it is important to protect your working surface, floor and clothing. Spread several layers of newspaper in the area where you will be working. Now place a pad of folded newspaper in the centre. On top of this, place a thin piece of hardboard, and then cover with a damp kitchen cloth. Smooth the cloth flat to stop marks being transferred to the finished paper.

6 Give the pulp another good stir. Place the deckle on top of the mould, with the net side uppermost and lining up the edges exactly. Grip the shorter edges, pressing the pieces firmly together. The deckle will stop the pulp running off the mould and give the paper neat straight edges.

7 Push the mould and deckle slowly, and at an angle, into the pulp working from the far side of the vat. As you push further into the pulp, so more pulp will be collected on to the net. If you have trouble collecting the pulp on the net, the ratio of pulp to water may be wrong and will need adjusting. Too much pulp and the paper will be thick and lumpy; too thin and it will be thin and holey.

8 Straighten up the mould and deckle so that the net is just below the surface of the pulp. Gently lift the mould and deckle to check that you have an even thickness of pulp on the net. If you are unhappy with the spread of pulp, scrape it back into the vat and start again.

9 Lift the mould and deckle right out of the pulp, and then gently rock it backwards and forwards and from side to side. This will help to settle the fibres and give a flat even sheet of paper. Do not overdo the rocking or the pulp will become holey. Allow the excess water to drain away; this can take several minutes, so place the mould and deckle across one corner of the vat.

10 Take away the deckle from the top of the mould. Some water may have been caught around the inside edges of the deckle; allow this to drain, keeping the mould flat until the water has gone. Turn the mould over so that the newly made sheet of paper faces downwards.

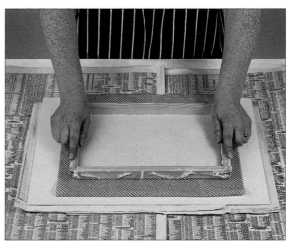

11 Although the paper will be held reasonable firmly on the net, it is advisable to transfer the paper to the kitchen cloth quickly. If the paper does fall off the mould, scrape it back into the vat and make the sheet again. In one gentle rolling movement, place the mould on to the cloth, press down on one short edge, and lift up on the opposite edge.

12 Roll the mould up and away, leaving the sheet of paper on the kitchen cloth. You may need a few goes to get an even layer of paper on the net, and to get the paper to stay on the kitchen cloth when you roll the mould away. If things go wrong put the pulp back into the vat and start again.

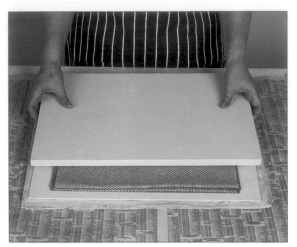

13 Place a damp kitchen cloth on top of the sheet of paper. Smooth the cloth removing any wrinkles and making sure the surface is perfectly flat. Make four or five more sheets, and place them on top of the first with a damp kitchen cloth between each. Cover the final sheet with a damp cloth, and then cover the stack with a clean piece of hardboard.

15 Place a sheet of plastic on your work surface. Remove the board and the cloth from the top of the stack. Carefully take off a kitchen cloth with its sheet of paper, and lay it on the plastic sheet to dry. The paper will be very strong at this stage, so do not worry too much about damage. Depending on the thickness of pulp the paper may take three or four days to dry.

14 Place two clean bricks on top of the stack and leave for several hours until quite a lot of the excess water has been soaked up into the cloth or has run out on to the paper. This process is called 'couching', and is the point in paper making where you stack, press and dry the newly made paper.

16 If you want the paper to have a smooth surface, you will need to roll it flat while slightly damp. Use a rolling pin to give the paper a flat smooth surface, but do not press too hard or the paper may tear. Once pressed, dry the sheets separately under heavy weights.

Making Decorative Paper

Once you have mastered the art of making plain paper try adding decoration using things found in your kitchen cupboards or around the house: fruit juice, thread and material scraps, coloured napkins, food colouring and fruit and vegetable peelings, all make wonderfully textured and coloured paper at very little cost

Making floral paper

Drop dried petals and leaves on to the surface of the pulp before making each sheet of paper; or to give the paper more colour, stir crushed potpourri into the pulp.

Making Christmas paper

Cut paper napkins or coloured paper into small pieces and stir it in the pulp: the dye will come out and colour the pulp. Add tinsel and glitter for a sparkley Christmas effect.

Making scrap thread paper

Cut material, ribbon and thread into small pieces and sprinkle it on the surface of the pulp. Try adding ribbon roses, metal foil and dried pulses for a more three dimensional look.

Making coloured paper

Add food colouring, spices, instant coffee, onion skins boiled in water, and fruit tea to the pulp to make pale coloured paper. For a brighter colour use poster paint.

Making fruit coloured paper

Fruit juice can be used to colour pulp and will give a reasonably strong result. Bring soft fruit like blackberries, blackcurrants and raspberries to the boil in a small amount of water. Simmer for a few minutes and then remove from the heat; allow to cool. Strain off the juice and stir into the pulp.

Making vegetable paper

Dissolve washing soda in a pan of water; add chopped up plant material. Bring the water to the boil and then simmer for three quarters of an hour. Drain the pulp through a net, and then rinse under the cold tap. Beat to a smooth, creamy pulp, then make a sheet of paper following the instructions on page 8.

Making paste paper

Paint a generous layer of wallpaper paste over the surface of a sheet of paper. Apply several colours of acrylic paint to the paste, then blend the colours together using a soft paintbrush. Use the blunt end of a paintbrush to draw patterns in the wet paste. Dry over the slats of an airing cupboard.

Making fold 'n' dye paper

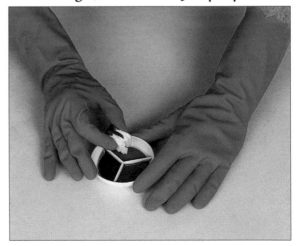

Concertina fold a sheet of Japanese paper. Dip the sides of the folded paper into food colouring, ink or silk paint: the dye will create patterns on the paper. Cover the damp paper with a sheet of clean paper and press to remove the excess dye. Dry flat, then cover with clean paper and iron on a low heat.

Making batik paper

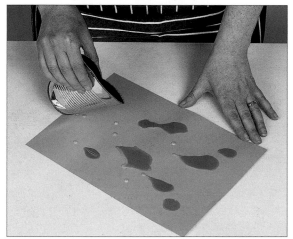

1 Melt batik wax in the top of a double boiler. Put the wax in an empty pet food tin, squashing the sides together to form a spout. Make patterns by pouring the wax on to a sheet of pale coloured paper; or white paper painted with watery poster paint. As the wax touches the paper it will dry almost immediately.

2 Paint the batiked paper with watery paint using long brush strokes: the paint will not stick to the wax. Once the paper is dry, sandwich it between newspaper. Turn the iron to a cotton setting and iron the paper – the melted wax will be absorbed into the newspaper leaving a pattern.

Making marbled paper

1 Three quarters fill a vat with wallpaper paste and leave it to stand until the paste reaches room temperature. Using a teaspoon, drop small amounts of different coloured paint on to the surface of the paste – you can use marbling paints or oils mixed with turpentine. Swirl the paint around, taking care that it stays on top of the paste.

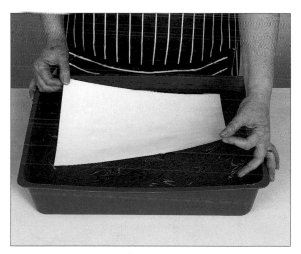

2 Lower a sheet of paper very gently on to the surface of the paste – this will stop air bubbles forming on the surface. Take the paper out of the vat and lay it on newspaper. Hold the marbled paper under cold running water, preferably a shower spray. Dry on a washing line, then flatten under weights.

Marbled Daisy Frame

Creating beautiful marbled paper is very easy to achieve using oil paints and wallpaper paste. Each sheet of paper that you make will be unique, mirroring the patterns created when paint is swirled in a vat of wallpaper paste. Once dry, the paper has been used to cover a picture frame and hexagonal box

You will need

- Flat MDF Frame 30x40cm (12x16in)
- Hexagonal shaped box
- Paper (good quality drawing paper)
- Paper for making pulp
- Oil paints or marbling paints –
  rose, flesh, magenta, blue
- Turpentine, wallpaper paste
- Light coloured raffia, mount board
- Plastic tray to use as a vat and
  marbling tray
- Kitchen cloths, hardboard rectangles, weights
- Beetroot – cooked (not in vinegar)
- Food colourings – yellow and green
- Gold rubbing paste
- Teaspoon or pipette, thin wire (coat hanger)
- Mould and deckle, newspaper
- Bucket, length of wood or hand held liquidizer
- Tacky glue, masking tape, paintbrushes
- Craft knife, pencil, ruler, scissors, water jar

Preparing the marbling tray

1 Before beginning, protect your work surface with several layers of newspaper as oil paints can mark surfaces and clothing.

2 Mix the wallpaper paste, following the instructions on the packet for making 'size'. Make enough paste to three quarters fill your vat. Leave to stand until the paste reaches room temperature.

3 Cut ten pieces of drawing paper to fit inside the tray; or use handmade paper (see Making a Sheet of Paper, page 8).

4 To marble the paper you can use oil or specialist marbling paints: oil paints are messier to use than marbling paints, but give a brighter, sharper colour. The oils will need to be mixed with a little turpentine to make the paint more liquid.

5 Using a teaspoon or a pipette, add equal drops of rose and flesh coloured paint to the surface of the paste. In a saucer mix magenta with a little flesh and add a small amount to the paste. Finally, add a tiny spot of blue. The colour should spread about 2.5cm (1in) across the paste. If the paint does not spread then thin it with a little more turpentine; if it spreads too far and too thinly, add more undiluted paint; if the colours sink, the paste

may be too thick, and needs to be thinned with a little water.

6 Using a thin piece of wire, gently swirl the paint around in the tray until you are satisfied with the design. Do this very carefully so that the paint stays on top of the paste.

Making the marbled paper

1 Hold one of the pieces of paper that you cut to size by diagonal corners. Lower the paper on to the surface of the vat belly first, and then let the corners gently down in a rolling motion. This will prevent air bubbles getting trapped under the paper, which will leave white marks in the marbled design.

2 The paper will soak up the paint almost immediately. Take the paper out of the vat, holding it by the two corners on the long side, and lay it colour side up on newspaper.

3 Take the marbled paper to a sink or bath and run it under cold running water,

preferably a shower spray. Dry the paper flat on newspaper or hang it on a washing line: it may take two or three days to dry. Flatten the paper under heavy weights before using.

4 Before making the next sheet of paper, skim any remaining paint off the surface of the paste, and then add more paint as before.

5 Make enough paper with a similar pattern to cover the frame and box.

Covering the frame

1 Lay the paper on the frame; if the paper is large enough you will be able to cover the frame in four pieces, with a mitre at each corner. If you have to cover each side using several strips of paper, make sure the overlaps are on on the diagonal. Glue the paper on to the frame, smoothing out any air bubbles. Once the glue is dry, cover the reverse side of the frame in the same way.

Making paper

1 Make ten sheets of natural paper (see Making a Sheet of Paper, page 8). Recycle only light coloured paper with long fibres, and then add raffia to the pulp to give it texture. The raffia should be cut into short lengths and then torn into shreds, before being added to the surface of the pulp for each sheet of paper (see Making Decorated Paper, page 13).

Making the daisy picture

1 Cut a piece of mount board to fit snugly within the hole in the picture frame: this will be the back of the daisy picture. Cut a piece of natural handmade paper the same size as the board, and glue it in place.

2 Tear a rectangle of handmade paper smaller than the backing board, but large enough

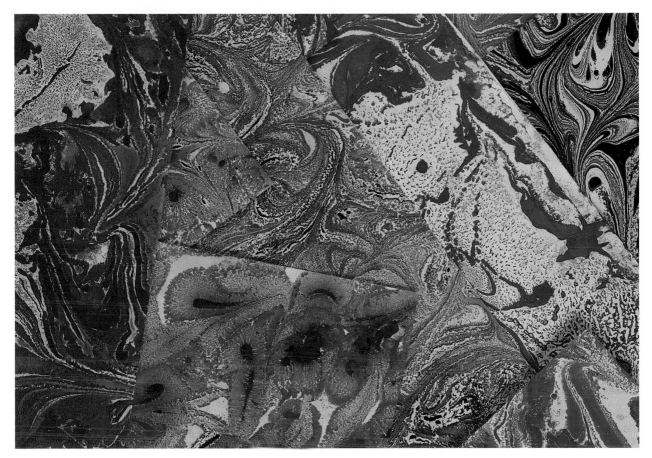

to fit behind the daisy picture. To make the paper easier to tear, fold the paper where you would like the tear to be and then dampen it with water before tearing. Do not use scissors or a knife to cut the paper or you will not create an uneven edge.

3 Extract the juice from the cooked beetroot by cutting it up and allowing the juice to drain down through a strainer.

4 Place the torn rectangle of handmade paper on a piece of scrap paper and then dampen it with water. Using a teaspoon or pipette, apply drops of beetroot juice to the surface of the paper working from the middle. The damp paper will soak up the juice fairly evenly, but you may need to keep adding juice to achieve a strong colour. The strongest colour should be

in the middle of the paper, fading to a paler pink at the edges. Once dry, glue the paper to the centre of the covered backing board.

5 Make tracings of the daisies, leaves and stem on page 21. Cut out the tracings and lay them on to natural coloured paper. Draw around the edges, transferring the designs on to the paper. Cut out the shapes.

6 Dampen the centres of the daisies, leaves and stem with water. Apply a few drops of yellow food colouring to the centre of the daisies, and green and yellow to the leaves and stem.

7 Once the paper is dry, apply gold rubbing paste with an old paintbrush or cotton bud to the edges of the design, and to add detail to the flowers, stem and leaves.

8 Arrange the stem, leaves and daisies on to the stained paper, and glue in place.

9 Put the completed picture in to the frame and secure with masking tape.

Covering the hexagonal box

1 Make enough marbled paper to cover the hexagonal box and lid. Cut a separate strip for each side of the box, allowing enough to turn over at the top and bottom. Cut a big enough piece of paper to cover the box top and sides, and extra to turn on to the inside. Apply a thin layer of glue to the box and then attach the paper sides, smoothing them down to remove any air bubbles. Glue paper on to the lid, cutting it up to the box at each corner. Turn the excess over, and glue on to the inside of the lid. Cover the bottom, and the inside of the box and lid.

Making the daisy panels

1 Tear eight rectangles of natural coloured paper slightly smaller than the sides of the box – remember to fold the paper and dampen it before tearing. Once dry, glue the squares to the centre of each box side.

2 Tear eight more rectangles, slightly smaller than the first. Stain them with beetroot juice in the same way as before, and glue them in the centre of the natural coloured rectangles.

3 Make a tracing of the small daisy design on the opposite page, and use it to cut eight daisies from natural coloured paper. Colour the centres with yellow food colouring, and add detail at the edges and centre with gold rubbing paste. Glue a daisy on to the centre of each rectangle.

4 From natural coloured paper, tear a hexagon slightly smaller than the box lid, and glue

it on to the centre of the lid. Tear another hexagon slightly smaller than the first; dampen it with water and then colour with beetroot juice as before. When dry, glue it to the centre of the natural paper panel. Cut a daisy from natural coloured paper. Colour the centre and edges as before and then glue to the centre of the lid.

Making the tags

1 Cut two pieces of thin cream coloured card 7.5x8.5cm (3x3½in).

2 Tear two pieces of natural coloured paper, large enough to just overlap the edges of the card; glue one on to each card.

3 For the cream daisy tag, tear a piece of natural coloured paper 5x6cm (2x2½in). Colour with beetroot juice as before, and then glue to the centre of the tag.

4 Cut a small daisy from natural coloured paper, colouring the centre with yellow, and then adding detail to the edges with gold. Glue the daisy on to the tag.

5 For the larger daisy tag, make a tracing of the larger single daisy on the opposite page. Cut the daisy from natural coloured paper, and then colour with beetroot juice.

6 Tear a small circle of natural coloured paper to fit the centre of the daisy. Colour with yellow food colouring, and then add detail with gold rubbing paste. Glue the circle on to the daisy, and the daisy on to the tag.

7 Punch two holes in the middle of one short side of each tag. Cut two lengths of raffia 25cm (10in) long. Thread raffia twice through the holes in one of the tags, before tying the ends in a knot. Repeat for the other tag.

Large Daisy Tag

Box Top
and
Picture

Small Daisy Tag

Use these outlines to cut daisy shapes from
handmade paper

Plant Paper Pictures

It is possible to make paper from waste plant material, as long as it has a stringy fibrous texture, and is not too tough. The process is slightly more involved than when using paper waste, as the plant material must be broken down to release the fibres. Three plant papers were used for this project: celery, leek and pineapple top

Some plant material will turn into a black mushy pulp as soon as it is heated, and no amount of bleach will make it usable; others remain hard and fibrous however long they are boiled. If this happens, throw away the material, and start again. Making paper from plant material can be unpredictable, so trial and error is the only way to attempt making this paper.

You will need

- Plant material – pineapple tops, celery, leeks
- Paper for making pulp
- Two picture frames – 21.5x26.5cm (8½x10½in)
- Brown parcel tags, paper – green
- Washing soda, garden twine – green
- Powdered poster paint – green
- Food colouring – green
- Acrylic paint – light and dark green
- Beetroot juice, blackberry juice
- Large saucepan, jug, funnel, ruler, pencil
- Mould and deckle, plastic tray to use as vat
- Muslin or net, kitchen cloths, paintbrushes
- Piece of wood or hand held liquidizer
- Newspaper, rectangles of hardboard, scissors
- Clean bricks or heavy weights
- Bucket, water
- Teaspoon
- Tacky glue
- Rolling pin

Making plant paper

1 To make vegetable and fruit paper you will need to find plant material with long narrow leaves like daffodils, rushes, grasses, nettles, irises; stringy stalks like celery, rhubarb, cow parsley, sunflower, leek; and onion peelings, pineapple tops and cauliflower leaves. Remove any woody, tough stems from your plant material and then cut into small pieces.

2 Dissolve 30g (1oz) of washing soda in a jug of cold water and pour it into an old saucepan. Do not use an aluminium pan or the washing soda may damage the surface. Add more water until the pan is three-quarters full.

3 Put the chopped up plant material into the pan, making sure it is completely covered by the water. Bring the water slowly to the boil on the stove, then turn down the heat and let it simmer very slowly for about 45 minutes. Be careful not to let the pan boil dry.

4 After 45 minutes, check to see whether the plant material is ready for the next stage. Do this by removing a piece of plant material from the pan; allow it to cool for a few seconds then squeeze it between your finger and thumb. The plant material is ready when it feels soft and falls apart. Pineapple tops may take longer to soften than plants with soft stalks like rhubarb or celery.

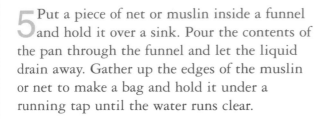

5 Put a piece of net or muslin inside a funnel and hold it over a sink. Pour the contents of the pan through the funnel and let the liquid drain away. Gather up the edges of the muslin or net to make a bag and hold it under a running tap until the water runs clear.

6 Washing soda can turn plant material a nasty dark brown colour. If this happens, dilute bleach with cold water in a bowl, then add the plant material to the water. Leave for an hour, before rinsing under cold water. The bleach will have lightened the plant material.

7 Beat the plant material to a smooth, creamy pulp, and then make paper following the same instructions for making paper using paper pulp on page 8.

Making vegetable pictures

1 You will also need to make two pieces of dark green paper coloured with poster paint (see Making Paper, page 13); two pieces of light green paper coloured with food colouring (see Making Paper, page 13); and one piece of natural coloured paper. You will also need a small amount of vegetable or fruit paper.

2 Take the backboards out of the frames. Cut dark green paper to the same size as the boards, and glue it in place. Cut light green paper slightly smaller than the inside edges of the picture frames. Give the edges a natural torn look, by folding and then dampening the paper with water before tearing on the fold. Glue the torn paper to the middle of the dark green paper.

3 To give the pictures a three-dimensional look, the leaves should be cut individually from different papers. To do this, make several tracings of the beetroot and radish designs opposite. Cut out the leaves individually;

where a leaf goes behind another, extend the tracing slightly to allow it to overlap. The beetroot is cut in one piece without the leaves; the radish in two pieces.

4 Place the individual tracings on to handmade paper, and draw around the edge. Cut out the pieces using sharp scissors. The roots should be cut from natural coloured paper, and the leaves from plant paper, using the lightest paper for the leaves at the back.

5 Dampen the beetroot with water. Using a teaspoon or pipette apply drops of beetroot juice to highlight the edges of the beetroot shapes, and where they overlap. Dampen and then highlight the radish roots in the same way using blackberry juice.

6 Unravel a few strands of green garden twine, and glue them to the underside of the beetroot at the points. Glue the beetroot on to one of the green covered backboards. Arrange the leaves and glue them in place.

7 Cut seven 4cm (1½in) lengths of green twine for the radish stems and glue them to the backs of the radishes. Glue the radishes and stems on to a piece of green covered backboard. Arrange the leaves and glue them in place.

8 Paint the frames with a coat of light green acrylic paint, and leave it to dry. Using a large paintbrush, drag dark green paint across the frame in the direction of the grain. When dry, secure the finished pictures in the frames.

Making gift tags

1 Using deckle edged scissors, cut squares of green card and glue them on to the parcel tags at an angle. Cut out leaf shapes from the plant paper and glue them in pairs on to the tags. Thread the tags with green twine.

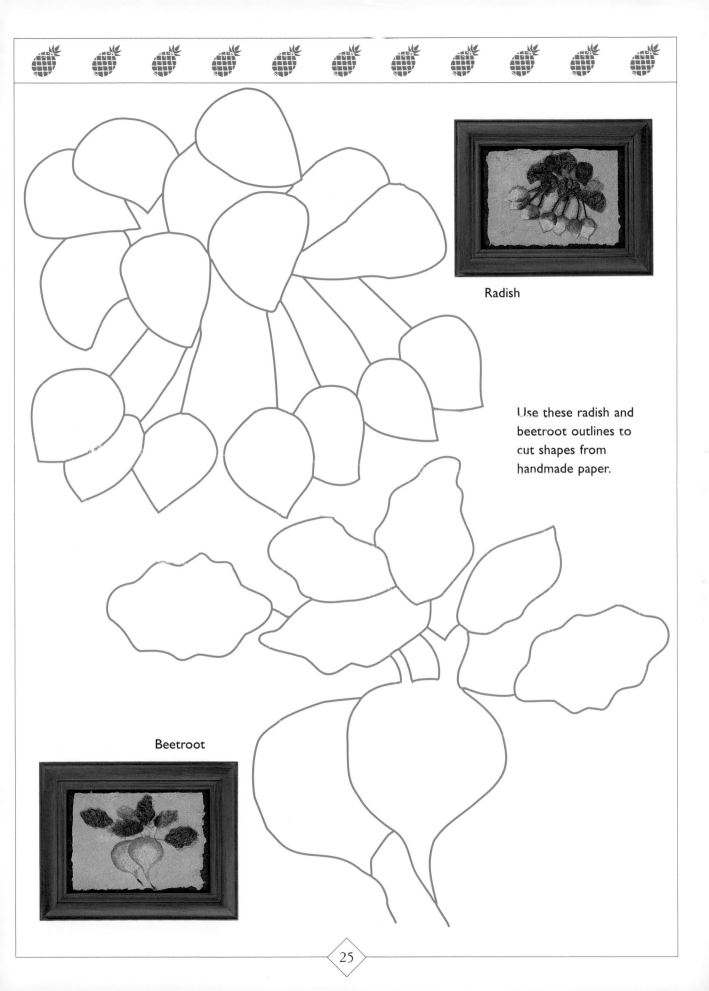

Radish

Use these radish and
beetroot outlines to
cut shapes from
handmade paper.

Beetroot

Paste Decorated Boxes

Bright geometric designs and fresh colours have turned these plain cardboard shoe boxes into a set of stylish storage boxes. Wallpaper paste and paint have been used to decorate the surface of the handmade paper, which is made from recycled envelopes and computer print-out paper

You will need

- Shoe boxes
- Paper for making pulp
- Wallpaper paste
- Acrylic paint – apple green, turquoise, lavender
- Food colouring – cerise
- Cream card and raffia for making tags
- Mould and deckle
- Bucket, length of wood
- Plastic tray to use as a vat
- Kitchen cloths, rectangle of hardboard
- Newspaper
- Clean bricks or heavy weights
- Mixing bowl, hole punch
- Paintbrushes – large and small
- Craft knife and cutting mat, scissors
- Tacky glue, water

Making paper pulp

1 Make a mould and deckle following the instructions on page 8. Recycle only light coloured paper for this project; computer print-out paper, paper bags and envelopes all have long fibres and will make strong paper. Avoid coloured or printed paper as you may find it difficult to cover the resulting surface colour with paint. Do not use paper with a shiny surface as it may be coated with clay, and will leave powdery patches on the finished paper (see Choosing the Right Paper, page 6).

2 Look over the paper to be recycled; discard any covered with glue and take out staples. Tear the paper into small squares, about the size of a postage stamp.

3 Put the torn paper in a bucket, cover with cold water and leave to soak for several days. You will need to top up the water as it gets drawn into the paper.

4 When the paper has been well soaked, pour away any water that remains. Using a hand held liquidizer or a piece of wood, beat the paper to a mushy pulp. This will take quite a long time as the pulp needs to be very smooth and creamy.

5 Half fill a plastic tray with water. Pour pulp into the tray so that there is a ratio of

approximately one third pulp to two thirds water. Stir the pulp and water together.

6 The wet sheets of paper will hold a lot of water, so it is important to cover your work surface with newspaper. Place a folded pad of newspaper in the centre, and on top of this a piece of hardboard and a wet kitchen cloth.

7 Give the pulp another good stir. Place the deckle on top of the mould, with the net side uppermost and lining up the edges exactly. Press the pieces firmly together.

Making a sheet of paper

1 Push the mould and deckle into the pulp (see Making a Sheet of Paper, page 8). Straighten up the mould and deckle just below the surface of the pulp. Gently lift, letting the excess water drain back into the vat. Remove the deckle from the mould.

2 Turn the mould over so that the paper is upside down under the mould: the wet net will hold the paper firmly in place. In one gentle movement, transfer the wet sheet of paper to the kitchen cloth. Press down on one short edge of the mould, and then lift up the opposite leaving the paper on the cloth.

3 Make a stack of five sheets of paper, laying a wet kitchen cloth between each one. Cover the stack with a heavy weight for several hours to press out some of the water.

4 Cover your work surface with a sheet of plastic, and then a layer of newspaper. Remove each kitchen cloth with its sheet of paper from the stack and spread them out over the protected surface. Leave until almost dry, this will take several hours.

5 To give the paper a smooth surface, roll each sheet with a rolling pin while it is still damp. To keep the paper flat, dry the sheets separately under a weighted board. Make enough sheets to cover your set of boxes.

Decorating the paper

1 Mix wallpaper paste with water to the consistency recommended for hanging lightweight wallpaper. Leave the paste to set overnight. Cover your working area with a thick layer of clean newspaper.

2 Lay a sheet of handmade paper on to the newspaper and paint a generous layer of paste over the sheet.

3 Apply apple green and turquoise or lavender and turquoise acrylic paint in random blobs onto the paste. Blend the colours together where they meet using a soft paintbrush.

4 Make diamond shapes over the surface of the paper, using the blunt end of a paintbrush to draw lines in the wet paste. Using the designs opposite as a guide, draw freehand cherry and pear shapes within the larger diamonds, and swirls in the smaller. Do not press too hard or the paper may tear.

5 Carefully hang each sheet of paper up to dry, laying the sheets on to a clothes horse or over the slats in an airing cupboard. Depending on the thickness of the paste, the sheets will take between 4-6 hours to dry. Paint enough sheets to cover your boxes.

6 If the paper has wrinkled, flatten each sheet under a heavy weight, and leave overnight.

Decorating the paper

1 Apply a thin, even layer of tacky glue to the outer surface of the box. Cover the box using sheets of decorated paper, smoothing each one flat to remove air bubbles. If the box is larger than the size of the paper, use several sheets, overlapping the edges where they meet.

2 Line the inside of the box with bright pink handmade paper. Refer to the instructions for making coloured paper on page 13.

Making a tag

1 Tags can be made using small scraps of decorated paper, glued to a rectangle of cream card. Punch two evenly spaced holes at the top edge of the tag. Thread with natural coloured raffia, and then attach to the box. Write the contents of the box on the tag, for easy reference.

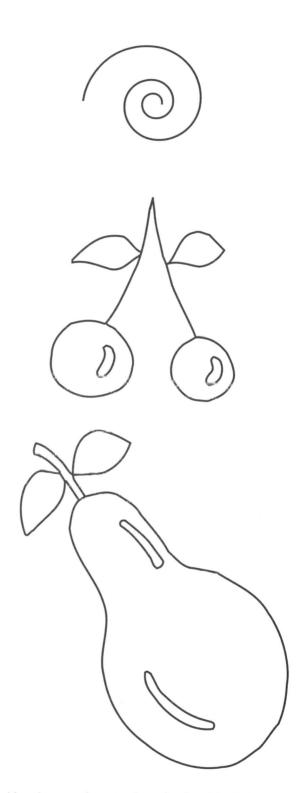

Use these outlines to draw freehand fruit shapes on to your boxes.

Tinted Paper Lampshades

Dark coloured spices like chilli and curry powder, and juice extracted from fruit can be used to colour paper. Although the ingredients may have a strong colour, once the pulp has been made into paper it will have just a hint of the original shade. The paper can be made into lampshades if lined with a fireproof backing material

Always use a fireproof backing material to line the shades if you are using them over an electric light bulb: do not use the shades over a lighted candle.

You will need

- Lampshade frames 8cm (3¼in) deep, 6.5cm (2½in) diameter at the top and 12.5cm (5in) at the bottom
- Fireproof sticky-backed lining for lampshades
- Chilli powder, blackberries
- Saucepan, muslin or fine sieve
- Cord, raffia, needle and thread
- Paper for making pulp
- Mould and deckle
- Plastic tray to use as vat
- Clean bricks or heavy weights
- Bucket, length of wood or hand held liquidizer
- Scissors, hole punch, rolling pin
- Newspaper, kitchen cloths, water
- Rectangles of hardboard
- Paper clips or clothes pegs, plastic bucket
- Tacky glue, craft knife and ruler

Making chilli paper

1 Prepare the paper pulp in the same way as for the Paste Decorated Paper on page 26. Pour the pulp and water into the vat. Sprinkle chilli powder on to the pulp mixture and stir well. Keep adding more chilli powder until you are happy with the colour. Use the pulp quickly as the chilli powder will give the pulp a very strong aroma. Make three sheets of chilli coloured paper.

Making the chilli lampshade

1 Measure the depth of the lampshade frame adding extra at the top and bottom for overlap. Cut a rectangle from each sheet of paper to this depth, and right across the width of the paper: the three sheets, once joined together and folded, will wrap around the frame. If you are working on a large frame, you may need to join in extra rectangles of paper.

2 Draw faint pencil lines every 2cm (³⁄₄in), in columns, on alternate sides of the paper. Using the back of a craft knife and a straight edge, score along each of the pencil lines. Fold the paper rectangles into a concertina.

3 Join the three sections together into one long rectangle by gluing the end pleats on each piece together: you may have to cut off some pleats, to allow for the concertina to fold correctly. Do not join the paper into a circle.

4 Flatten out the concertina folded rectangle of paper, and then cut fireproof sticky-backed lining material to the same size. Carefully peel off the protective backing and press the lining material firmly, sticky side down, on to the paper.

5 Score along the fold lines again, and then press the backed paper in concertina folds.

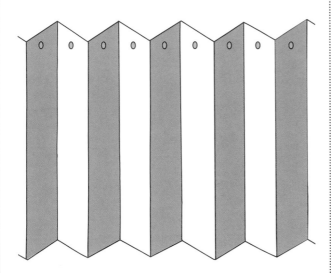

6 Using the hole punch, make a single hole in the centre of each column, 1cm (³/₈in) down from the top edge of the shade.

7 Thread the cord through the holes in the paper and then draw up the ends, so that the paper folds into a circle. Place the lampshade on to the frame and then glue the final two pleats together to complete the lampshade. Tie the cord into a bow.

8 Working from inside the frame, sew the cord on to the frame at 2.5cm (1in) intervals. This will hold the shade securely in place.

Making blackberry paper

1 Boil the blackberries in a small amount of water; leave them to cool and then strain the blackberry pulp through muslin or a fine sieve. Keep the juice.

2 Prepare the paper pulp in the same way as for the Paste Decorated Paper on page 26. Pour the pulp and water into the vat. Add the blackberry juice and stir well to disperse the colour. Make two or three sheets of blackberry coloured paper.

Making the round lampshade

1 If your lampshade frame has a depth of 8cm (3¹/₄in), a top diameter of 6.5cm (2¹/₂in) and a bottom diameter of 12.5cm (5in), then make a tracing of the lampshade design on the opposite page. To make your own tracing, place your frame on to a piece of white paper and then slowly roll it across the surface of the paper, marking the outline with a pencil. Add a small amount at both ends for joining. Cut out the tracing and place it on to the frame. Make adjustments until you are happy with the fit.

2 Lay the lampshade tracing on to the blackberry coloured paper, and cut along the pencil lines. You may need to join several sheets of paper together: make the joins part of the design by tearing the edges, rather than cutting them (see Glittery Christmas Paper, page 40). Line the shade with the fireproof backing in the same way as for the chilli paper.

3 With the lampshade still flat, punch holes at regular intervals along the bottom edge. Thread raffia through the holes, securing the ends with glue. Glue a raffia bow to the top.

4 Glue the lampshade into a circle, holding it together until dry with paper clips or clothes pegs.

5 Working from inside the lampshade, sew the raffia on to the frame to hold it in place.

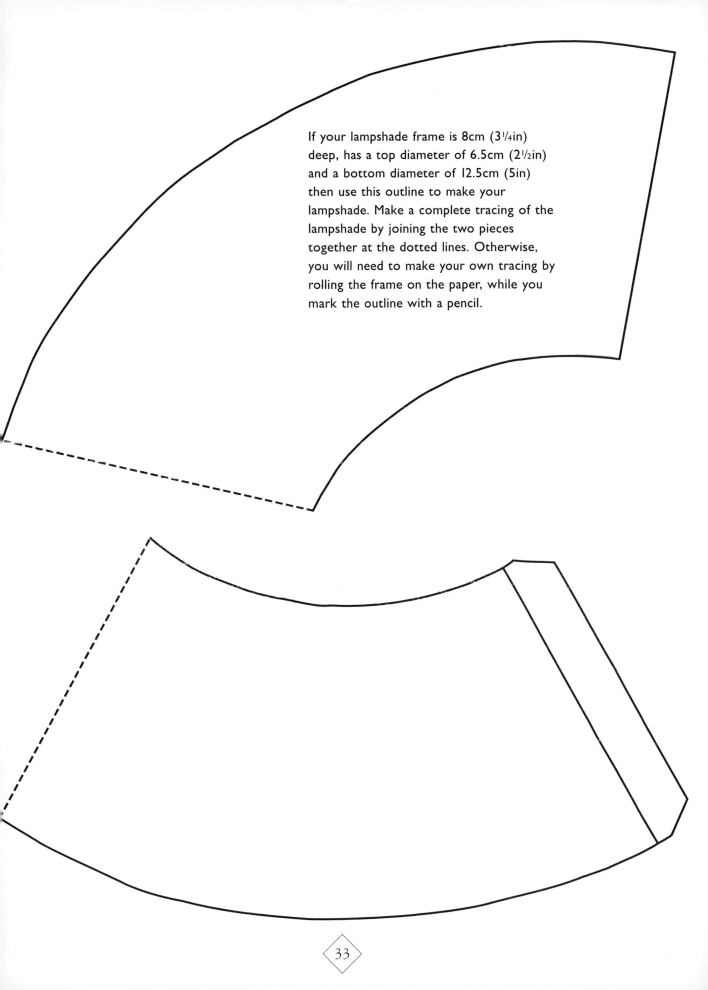

If your lampshade frame is 8cm (3¼in) deep, has a top diameter of 6.5cm (2½in) and a bottom diameter of 12.5cm (5in) then use this outline to make your lampshade. Make a complete tracing of the lampshade by joining the two pieces together at the dotted lines. Otherwise, you will need to make your own tracing by rolling the frame on the paper, while you mark the outline with a pencil.

Black Cat Desk Set

Food colouring is a very effective way to colour handmade paper. Small pieces of cream and white knitting and embroidery wool have also been added to the paper pulp, giving the paper texture. Once the folder and pen holder have been covered with the paper, they can be decorated with geometric borders and black cats

You will need

- File
- Note pad or pad of paper and pencil
- Stiff cardboard
- Empty cylindrical container
- 40cm (16in) black ribbon
- Paper for making pulp
- Food colouring – lavender, black
- Knitting and embroidery wool – cream, white
- Mould and deckle
- Bucket, length of wood or hand held liquidizer
- Plastic tray to use as a vat
- Kitchen cloths, rectangles of hardboard
- Clean bricks or heavy weights
- Masking tape, pencil, scissors, ruler, white paper
- Newspaper
- Rolling pin
- Craft knife and cutting mat
- Tacky glue, water

Making coloured paper

1 Prepare the paper pulp (see Making a Sheet of Paper, page 8). Pour the pulp and the water in to the vat and add a few drops of lavender food colouring, stirring well to disperse the colour thoroughly (see Decorating Paper, page 13).

2 Cut short lengths of cream and white knitting or embroidery wool and unravel some of the fibres. Give the pulp another good stir and then sprinkle the wool lengths onto the surface of the pulp.

3 Make twelve sheets of lavender coloured paper (see Decorated Paper on page 13).

4 As you make the sheets of paper you will need to keep the vat topped up with water, pulp and colouring. Try to keep the ratio of pulp to water the same as when you started. Stir the pulp between each sheet, adding more strands of wool to the surface.

5 Make two stacks of six sheets, separating each with a damp kitchen cloth. Cover each stack with a piece of hardboard, and weigh down with bricks or heavy books to press out some of the water.

6 Cover your working surface with a sheet of plastic, and then a layer of newspaper.

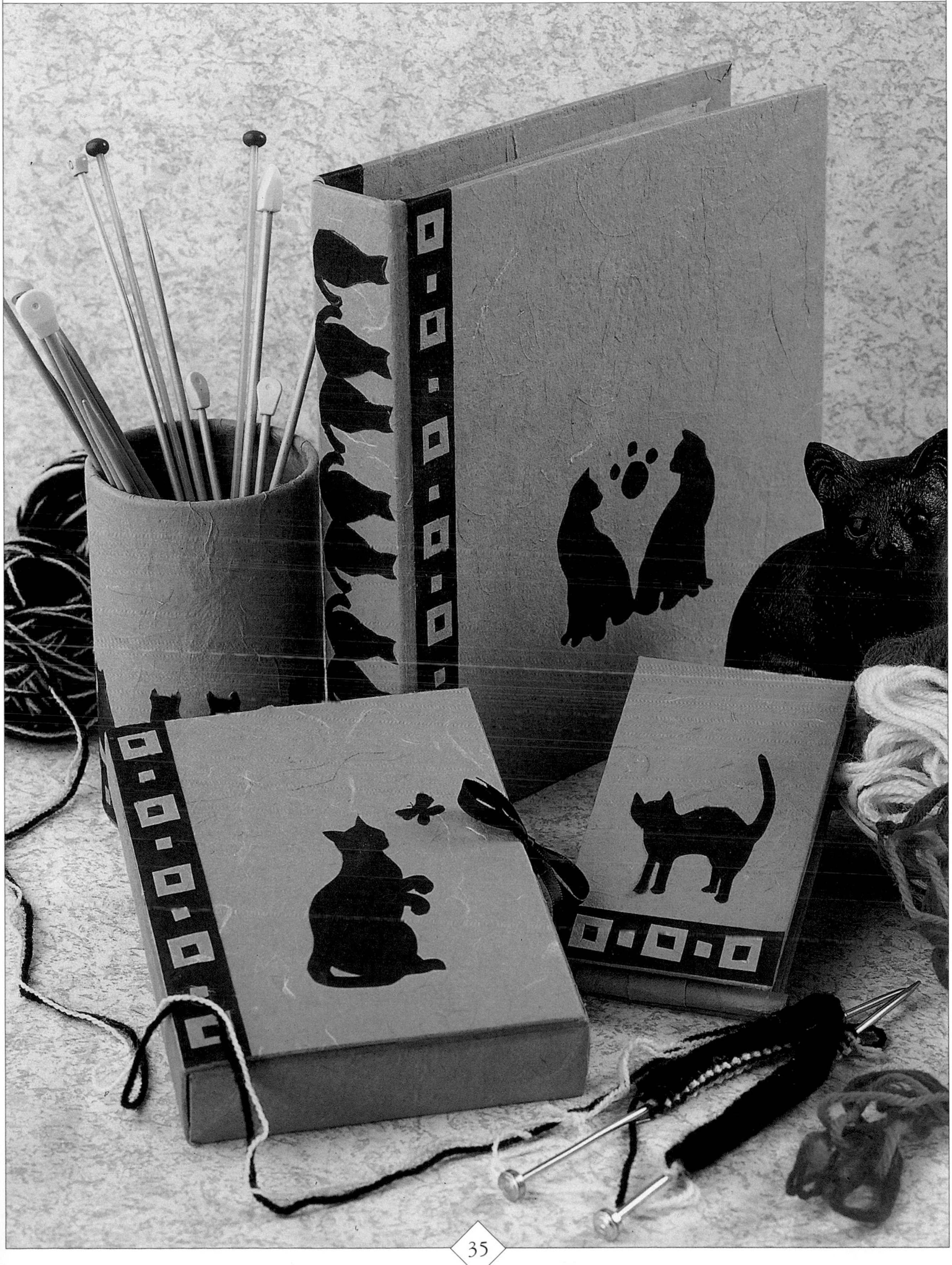

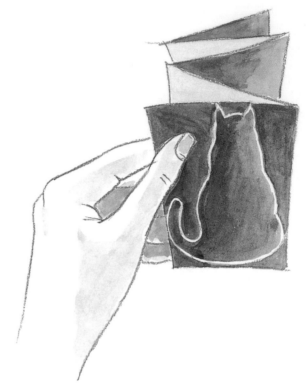

Remove each kitchen cloth with its sheet of paper from the stack and lay them on the protected surface. Leave until almost dry.

7 To give the paper an even surface, roll each sheet with a rolling pin while it is still damp. To keep the paper flat, dry each sheet separately under a weighted board.

8 Make four sheets of black paper using black food colouring.

Covering the pencil pot

1 Measure the circumference and height of the container you will be using for the pencil pot – use a cardboard cylinder container or clean food container. Cut a rectangle of lavender paper to fit around the outside, slightly taller then the container.

2 Apply glue evenly over the outside surface of the container. Wrap the paper around the container, lining it up with the bottom. Smooth out any air bubbles. Apply glue around the inside edge of the container and fold the paper in on to the glue.

3 To cover the inside of the cylinder, measure the height and circumference inside the container and cut the paper to fit. Apply glue over the surface of the paper, and press in position.

4 Cut a strip of black paper 2cm (³⁄₄in) wide and long enough to fit around the bottom of the container. Glue the strip in place.

5 Cut squares of lavender paper 1.25x1.25cm (¹⁄₂x¹⁄₂in). Cut out a small square from the centre of each one using a sharp craft knife and cutting mat. Glue the large and small squares alternately, on to the black band around the bottom of the container.

6 Make a tracing of the small cat design on page 39 on to white paper. Cut a strip of black paper 4cm (1¹⁄₂in), and long enough to go round the container. Place the tracing on to the left hand end of the black paper strip and draw over the design lines with a ball-point pen, making indentations in the black paper. Concertina fold the paper to the width of the cat, keeping the cat outline on the top. Make sure the folds are exactly aligned to the edges of the cat. Cut out the cat shape leaving two points uncut as shown on the diagram on page 39. Open out the string of cats, and glue them above the border around the container.

Covering the note pad

1 If your note pad has a hard outer surface, cover it with the lavender paper. Otherwise, make a hard cover for the paper pad by cutting two pieces of thick card slightly wider and 6mm (¹⁄₄in) longer than pad. For the spine, cut a piece of card the same width as the front and back, and the thickness of the pad.

2 Line up the three pieces of card with the spine in the centre, leaving a small gap between each to allow for them to fold. Join the pieces together by laying strips of masking tape between the spine and the cover pieces, holding them together. Repeat on the reverse side of the cover.

3 Cut one piece of lavender paper slightly larger than the complete cover. Glue the paper in place, folding the edges over and gluing them on to the reverse side. Make a scored indentation in the paper along both fold lines on either side of the spine.

4 Cut a piece of lavender paper for the reverse side of the cover. Glue this in place and then make scored indentations along the folds.

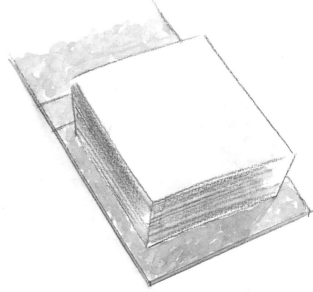

5 Spread glue over the back surface of the paper pad and stick it inside the hard cover.

6 Using a sharp craft knife or small saw, cut a pencil to the same width as the note pad. Cover the pencil with lavender paper. Cut a strip of lavender paper 6x2cm (2½x¾in) and glue the two ends together, making a loop big enough to hold the pencil. With the pencil in the loop, glue the ends of the loop to the inside back cover, below the paper pad.

7 Cut a strip of black paper 2cm (¾in) wide and the width of the cover. Glue this to the bottom edge of front cover. Decorate with lavender squares in the same way as for the pencil pot. Transfer the cat design opposite on to the black paper and glue it to the front cover of the note pad.

Covering the file

1 Cover the file with lavender coloured paper, allowing extra around the edges to turn over on to the inside – depending on the size of the paper you are making, you may need to overlap the paper. Cut lavender paper to fit the inside covers of the file, and glue in place. Make a scored indentation either side of the spine so that the file will fold.

2 Cut two strips of black paper 2cm (¾in) wide and the height of the file, allowing a little extra to turn in at the top and bottom. Glue the strips to the front and back cover either side of the spine. Cut lavender squares as before, and glue to the black strips.

3 Cut the two cats and the paw print design from black paper and glue them on to the front cover.

4 Cut a length of black paper 4cm (1½in) wide and the length of the spine. Make a tracing of the small single cat on to black paper, and concertina fold in the same way as for the pencil pot. Cut out the cat, leaving the

two joining points. Unfold the cats and glue them along the spine.

Making the notelet folder

1 Using the diagram and measurements on the opposite page as a guide, cut two rectangles from thick cardboard. Score along the dotted lines. Fold along the scored lines and then glue the flaps on the larger rectangle onto the smaller. Use masking tape to strengthen the joins on the inside of the folder.

2 Cover the outside of the folder with lavender coloured paper.

3 Cut a 16cm (6½in) length of black ribbon and glue one end to the inside middle of the flap against the outer edge. Glue lavender paper over the inside flap, covering the glued ribbon end.

4 Cut a 16cm (6½in) length of black ribbon. Using a sharp craft knife, make a small hole in the back of the folder, 1.5cm (⅝in) from the outer edge and lining up with the ribbon at the front. Push one ribbon end through the cut, and glue inside the folder. The ribbon ends can be tied together to hold the folder closed.

5 Cut a strip of black paper 2cm (¾in) wide and the height of the folder. Glue this against the spine on the front cover of the folder. Cut lavender paper squares as before, and glue them alternately along the black paper strip.

6 Make a tracing of the the cat and butterfly design opposite. Cut them from black paper and glue on to the front of the folder.

7 Fill the folder with folded note paper and envelopes, or notelets.

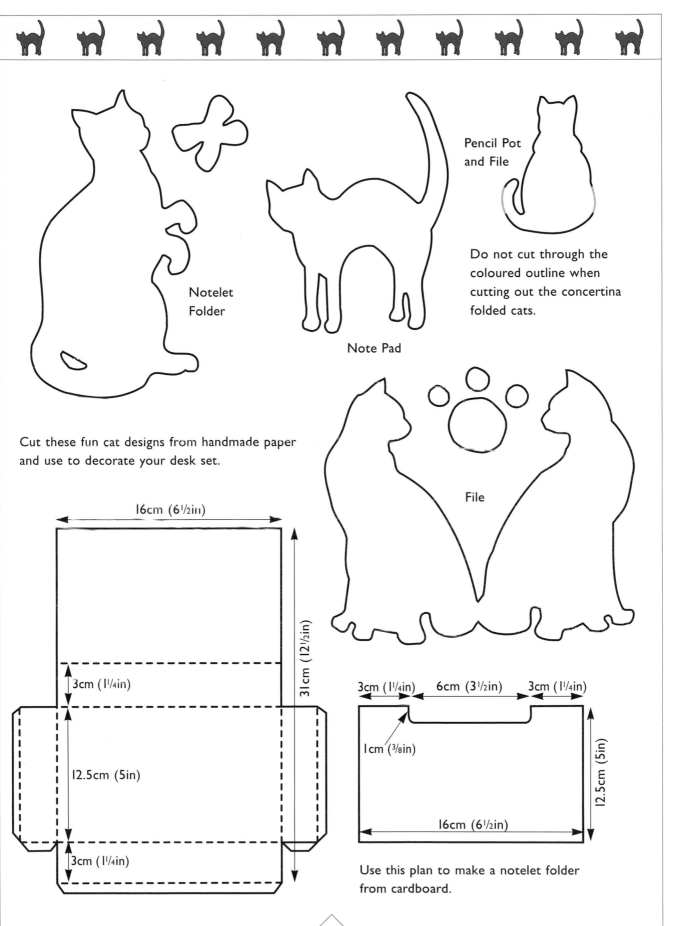

Pencil Pot and File

Do not cut through the coloured outline when cutting out the concertina folded cats.

Notelet Folder

Note Pad

Cut these fun cat designs from handmade paper and use to decorate your desk set.

File

16cm (6½in)

31cm (12½in)

3cm (1¼in)

12.5cm (5in)

3cm (1¼in)

3cm (1¼in)   6cm (3½in)   3cm (1¼in)

1cm (³⁄₈in)

12.5cm (5in)

16cm (6½in)

Use this plan to make a notelet folder from cardboard.

Glittery Christmas Cards

Paper napkins and tinsel have been added to the paper pulp before being made in to sheets of glittery paper. A stencilled star, Christmas tree and holly have been embossed and highlighted with gold rubbing paste, giving the cards, envelopes, tags and wrapping paper a co-ordinated and sophisticated look

You will need

- Envelopes and card – brown
- Paper napkins – red and green
- Paper for making pulp
- Bucket, piece of wood or hand held liquidizer
- Plastic tray to use as a vat
- Kitchen cloths, rectangles of hardboard
- Clean bricks or heavy weights
- Mould and deckle
- Rolling pin, sponge
- Star shaped stamp
- White writing paper
- Powder poster paints – red and green
- Tinsel – gold
- Metallic paint – gold
- Thin ribbon – gold
- Rubbing paste – gold
- Blank stencil card
- Light box or window, embossing tool
- Craft knife, cutting mat, pencil, ruler
- Newspaper, tacky glue, water

Making the paper

1 Prepare the paper pulp in the same way as for the Paste Decorated Paper on page 26. Pour the pulp and water into the vat. Tear red paper napkins into small squares and add them to the pulp; stir well. For a stronger colour, add red powdered poster paint to the pulp, again stirring well to mix in the powder. Make the green paper in the same way.

2 Remove some of the strands from a length of tinsel, and drop them on to the surface of the pulp just before making each sheet of paper (see Making Paper, page 13). For this project you will need five sheets of red paper and five sheets of green paper.

Making the holly card

1 Cut a piece of red handmade paper 18x25cm (7x10in) giving the edges a torn effect. To do this, fold the paper where you want the tear to be, dampen with water and then tear along the fold. Fold the paper exactly in half.

2 Make a tracing of the holly design on page 43. Transfer the design on to stencil card and then cut along the outlines, leaving leaf and berry shaped holes cut in the card.

3 Tear a piece of green paper 8.5x13cm (3¼x5in). Hold the stencil firmly on to the middle of the paper. Using a small piece of

sponge, dab gold rubbing paste on to the paper through the holes cut in the stencil. Leave the card to dry.

4 Place the paper, with the stencil still in position, on a light box (or against a window) so that the stencil is face down, and you are working from the reverse side of the design. Draw around the edges of the design, using an embossing tool, pressing as hard as you can without tearing the paper. Depending on the thickness of the paper it may not be possible to see through the paper, in which case 'feel' your way round the edges with the tool. Rub over the rest of the design as though colouring with a pencil, making the leaves and berries slightly concave from the back.

5 Carefully remove the stencil from the green paper. Glue the embossed green paper on to the centre front of the red card. To finish the card, cut a piece of white writing paper slightly smaller than the card; fold it in half and glue it inside the card along the spine.

6 To make the matching envelope, cut a piece of red paper 6x7cm (2½x2¾in) tearing the edges as before. Glue this to the top left hand corner of the envelope. Tear a piece of green paper 4.5x4.5cm (1¾x1¾in). Make a stencil of

the small holly leaf on the opposite page. Stencil and then emboss the holly leaf on to the paper. Glue it to the centre of the red paper on the envelope.

Making the tree card

1 Cut a piece of red handmade paper 18x25cm (7x10in) tearing the edges as before. Fold the paper in half.

2 Make a stencil of the Christmas tree design opposite. Cut a piece of green handmade paper slightly larger than the tree; stencil and then emboss the tree on to the paper. Cut round the gold embossed tree leaving a margin of about 3mm (⅛in) around the edges. Glue the tree on to the card. Glue a piece of folded white paper inside the card in the same way as for the holly card.

3 To make the matching envelope, tear a piece of red paper 5.5x5.5cm (2¼x2¼in) and glue it at an angle on the top left hand corner of the envelope. Stencil and then emboss the gold star from the top of the Christmas tree on to a piece of green handmade paper. Cut it out with a margin of 3mm (⅛in) and glue it to the centre of the red diamond.

Making the star card

1 Cut a piece of green handmade paper 18x25cm (7x10in), tearing the edges as before. Fold it in half. Tear a piece of red handmade paper 7.5x13.5cm (3x5½in) and glue it to the front of the green card.

2 Stencil and then emboss the star on to red handmade paper. Cut out the star leaving a small margin of red paper around the design. Cut out the centre of the star using a craft knife; put this small star to one side. Glue a piece of green paper behind the cut-out smaller star. Glue the star to the centre of the card.

3 Tear a 5x6cm (2x2½in) rectangle of green paper, and glue it to the top left corner of the envelope. Tear a red rectangle 3x4cm (1¼x1½in) and glue this to the centre of the green. Glue the small star, cut from the larger, on to the envelope.

Making the paper and tags

1 Paint a thin layer of gold paint on to the surface of your star shaped stamp, and press

it down firmly on to a piece of handmade paper. Add more paint to the stamp, and stamp the paper again. Repeat the process until the paper is covered with stars. Once dry, the paper can be used as gift wrap.

2 To make the tag, cut a rectangle from stamped paper and the same sized rectangle from card. Glue the two together; punch a hole in one corner and thread with gold ribbon.

Use these Christmas shapes to cut stencils for decorating your handmade paper.

Petal and Leaf Stationery

Not only will this stationery set look good, but it will also smell good. Potpourri and pressed flowers have been used as an ingredient in the paper making process, giving the paper texture as well as scent. To give this set an autumnal feel, finish with plaits and tassels made from thick embroidery thread

You will need

- Stationery box
- Address book
- Paper for making pulp
- Dried leaves and flowers
- Potpourri
- Thin card
- Cardboard tube
- Wool or thick embroidery thread
- Mould and deckle
- Bucket, length of wood or liquidizer
- Plastic tray to use as a vat
- Kitchen cloths, rectangle of hardboard
- Newspaper
- Clean bricks or heavy weights
- Craft knife and cutting mat, scissors
- Tacky glue, masking tape, water

Making coloured paper

1 Prepare the paper pulp (see Making a Sheet of Paper, page 8). Recycle only light coloured paper for this project; although beige coloured envelopes and paper bags can also be used, as long as the colour is not too strong (see Choosing the Right Paper, page 6).

2 For this project you will need to make three different types of paper: natural, potpourri textured, and some decorated with dried flowers and leaves.

3 Using the prepared paper pulp, make one sheet of natural coloured paper (see Making a Sheet of Paper, page 8).

4 Add more water and pulp to the vat, keeping the ratio of pulp to water the same as when you started. Give the pulp a good stir and then drop dried flower petals and leaves on to the surface of the pulp. You can dry your own petals and leaves by pressing them between the pages of a heavy book for several weeks, or they can be bought from a craft shop.

5 Make four sheets of petal and leaf paper (see Decorated Paper, page 13). Although you will not use all four sheets to cover the stationery box, you will be able to choose the parts of the paper with the best concentration of petals and leaves.

6 Remove any petals or leaves that are still on the surface of the pulp, and then top up the vat with pulp and water.

7 Put a good handful of potpourri in to a paper bag, seal the top and then crush the potpourri in to very small pieces.

8 Sprinkle crushed potpourri pieces into the vat, and give the pulp a good stir. Keep adding potpourri and stirring until you are happy that it is evenly spread throughout the pulp.

9 Make four sheets of paper using the potpourri pulp, in the same way as before.

Covering the stationery box

1 Buy a stationery box with a drawer and lift up lid. Remove the drawer from the box and cover the front with petal and leaf, and the sides with natural paper: use natural paper in places where the paper needs to be flatter.

2 Using wool or thick embroidery thread, make a short plait; knot one end together to stop it unravelling, and apply glue to the

other. Glue the plait to the underside of the drawer. Cover the base of the drawer with natural paper.

3 Make another plait to go between the lid and the top of the box. Glue the plait in place, and then cover the box in the same way as the drawer. Try to cut the paper so that the best petals and leaves are positioned prominently on the box.

4 Strengthen the join between the lid and the box with a strip of natural paper, glued between the two parts.

5 Put the drawer back in the box. Make two bundles of stationery and tie them up with lengths of wool. Place one bundle in the drawer, and the other in the recess at the top of the box.

Covering the address book

1 Cut a piece of potpourri paper slightly larger than the cover of the book. Glue the paper on to the front and back cover. Make a small cut in the excess paper either side of the spine, top and bottom; add glue to the flaps of

paper, and then push them down between the spine of the book. Glue the excess paper at the edges on to the inside of the covers.

2 Tear a leaf roughly out of the petal and leaf paper, and glue it to the front of the book. Make a long plait of wool or embroidery thread; loop it over the top of the book, and then glue the plait down the folds on the front and back, where the spine and cover meet. Knot the ends together at the bottom.

Making the bookmark

1 Cut two rectangles of thin card 4x15cm (1½x6in). Cover one side of each rectangle with potpourri paper.

2 Cut thirty 10cm (4in) lengths of wool to make the tassel. Fold these over a longer length of wool. Tie the bunch together, 4cm (1½in) from the folded ends.

3 Glue the rectangles of paper covered card together, with the tassel ends between the two cards, and the tassel protruding at one end of the bookmark. Glue a single strand of wool around the edge, pushing the ends up between the card layers. Decorate the bookmark with a torn leaf in the same way as for the book.

Making the basket

1 Cut 4cm (1½in) from an empty cardboard tube. Cut a circle of card for the base, and a card strip for the basket handle. Glue in place then cover the joins with masking tape

2 Cover the basket, inside and out, with petal and leaf paper, and then fill with potpourri.

Scrap Paper Photo Albums

This wonderfully textured paper is made by adding a variety of scrap material to the paper pulp: thread, fabric, ribbon and wool; beads, sequins and pulses like split peas and lentils make interesting effects. Here we have used the paper to cover photograph albums, which has made the covers as interesting as the photos inside

You will need

- A2 card – silver
- Small pieces of wool, ribbon, material, beads, lentils, small ribbon rosebuds
- Silver ribbon for the leporello ties
- Stiff card for the album covers
- Thin frosted plastic cut from a video cover
- Hole punch, two metal screwbinders
- Stiff paper for album pages and spacers – white
- Paper for making pulp
- Mould and deckle
- Bucket, length of wood or hand held liquidizer
- Newspaper, rectangles of hardboard
- Clean bricks or heavy weights
- Scissors, rolling pin, screwdriver
- Tacky glue, kitchen cloths, water
- Craft knife and cutting mat

Making the scrap paper

1 Choose light coloured paper with long fibres for this project (see Choosing the Right Paper, page 6). Tear the paper into small squares about the size of a postage stamp. Put the torn paper into a bucket, cover with cold water and leave to soak for several days (see Paste Decorated Paper, page 26). You will need to top up the water as it gets drawn into the paper.

2 When the paper has been well soaked, pour away any water that remains. Using a hand held liquidizer or a piece of wood, beat the paper until it is a smooth, creamy consistency.

3 Half fill a plastic tray with water. Put pulp into the tray so that there is a ratio of approximately one third pulp to two thirds water. Stir the pulp and water together.

4 Cover your work surface with plenty of newspaper. Place a folded pad of newspaper in the centre, and on top of this a piece of hardboard and a folded kitchen cloth.

5 Give the pulp another good stir. Place the deckle on top of the mould, with the net side uppermost and lining up the edges exactly.

6 Sprinkle small pieces of thread, wool, ribbon, material and beads on to the surface of the pulp in shades that work well

together. Ribbon roses and dried pulses like lentil can also be added to the pulp. Do not stir the pulp at this stage.

7 Holding the mould and deckle firmly together, push it into the pulp (see Making a Sheet of Paper, page 8), making sure you scoop up the threads and materials floating on the pulp. Straighten up the mould and deckle just below the surface of the pulp. Gently lift, letting the excess water drain back into the vat. Remove the deckle from the mould.

8 Turn the mould over so that the paper is upside down under the mould: the wet net will hold the paper firmly in place. In one gentle movement, transfer the wet sheet of paper on to the kitchen cloth. Press down on one short edge of the mould, and then lift up the opposite leaving the paper on the cloth.

9 Make a stack of at least three pieces of paper, laying a folded cloth between each one. Put heavy weights on the stack for several hours to squeeze out the excess water.

10 Cover your work surface with a sheet of plastic, and then a layer of newspaper. Remove each kitchen cloth with its sheet of paper from the stack and spread them out on the plastic. Leave the paper until almost dry, this may take several hours.

11 Roll each sheet of paper with a rolling pin to give it a nice even finish. Dry the individual sheets of paper under heavy weights.

Making the leporello

1 Cut three pieces of silver card 18x56cm (7¼x22in). Using a pencil and ruler, divide the strips equally into four sections. Use the back of a scalpel blade to score along your pencil lines. Concertina fold the strips along

these lines. Glue the strips together to make one long folded strip. Leave it to dry.

2 Cut two pieces of the handmade paper 13x17cm (5¼x6¾in). Cut two pieces of stiff card to the same size and glue the handmade paper to the card.

3 Cut two 45cm (17¾in) lengths of silver ribbon. Glue one length diagonally across the middle of the front cover of the leporello, and one across the back.

4 Glue one of the card backed pieces of handmade paper on to the front of the leporello, and one to the back. This will hold the ribbon firmly in place.

5 Fold up the leporello, tying the ribbon ends together to keep it closed. Press between heavy weights and leave to dry overnight.

Making the photo album

1 Choose two pieces of handmade paper for the front and back covers. Cut them to approximately 17x19cm (6¾x7½in). If you wish, you may tear the edges to give a natural

effect. To do this, first fold and then dampen the paper with water before tearing.

2 Cut two pieces of stiff card just smaller than the paper – these are the covers on to which the handmade paper is attached. On one cover draw a pencil line 2.5cm (1in) in from the left hand edge. Score along this line with a ruler and the back of a scalpel. This will form the hinge for the front cover of the album.

3 Glue the pieces of handmade paper to the card covers. Place the assembled covers between clean card and put under a heavy weight until the glue is dry.

4 Cut two pieces of the silver card 2.5x17cm (1x6¾in), and two of frosted plastic from a video cover. These pieces will cover the spine on the front and back cover, and protect them.

5 For the album pages, cut sheets of stiff white paper 16x18cm (6¼x7in). Score a line 2.5cm (1in) in from the shortest edge. Cut six pieces of stiff paper 2.5x16cm (1x6¼in) – these strips of paper will be placed between the pages of the album to act as spacers.

6 Punch holes in the spine pieces, covers and spacers, making sure that all the holes will line up when the album is assembled.

Assembling the album

1 Using a screwdriver, undo the two screwbinders – a screwbinder is made up of two metal parts that screw together, and extra metal spacers that can be added as you need to extend the pages in your album. Assemble the album, pushing the screwbinders through the two holes in the album parts. Start from the front with the frosted plastic spine, then the silver card spine, front cover, album pages with spacers between, back cover, silver card spine, and finally the plastic spine. Screw the back on the screwbinder.

Fold 'n' Dye Gift Bags

Japanese paper has very similar qualities to blotting paper: it is very absorbent and strong when wet. This makes it ideal to use in this project where the paper is folded, dyed and then folded again into gift bags and boxes. Food colouring, ink and silk paint can be used on Japanese paper, giving some amazing and very colourful results

For this project you will need to dye the following A4 sized sheets of Japanese paper: one for the pyramid box, one for the small bag, and two for the square box. For each larger bag, you will need to dye two A3 sheets.

You will need
- Japanese paper (or paper that is absorbent and strong when wet)
- Thin card or good quality paper
- Raffia ribbon – red
- Dye – food colouring, ink or silk paint
- Hole punch, pencil, ruler
- Rubber gloves
- Newspapers, white paper
- Printing roller or rolling pin
- Small containers for dye
- Clothes peg, elastic band
- Tacky glue
- Iron

Dying the paper

1 Concertina fold a sheet of Japanese paper from one of the short sides, so that you have a 2cm (³⁄₄in) wide strip of folded paper. Concertina fold the other way until you have a block of folded paper. Hold it together with an elastic band.

2 Cover your work surfaces with newspaper, and put on a pair of rubber gloves.

3 Food colouring, ink and silk paint can all be used to dye the Japanese paper. Put about 10ml of each of the colours that you will be using in to small containers.

4 Remove the elastic band from the concertina folded paper and, holding the paper tightly between your finger and thumb, dip the corners and edges into each of the coloured dyes: the dye will spread across the paper creating patterns (see Fold 'n' Dye Paper, page 14).

5 Place the dyed paper on to newspaper and partially unfold it. Cover the damp paper with a sheet of clean white paper and apply a little pressure with your hand, a roller or rolling pin – this will remove the excess dye and stop it from spreading further across the paper. Now completely unfold the paper and lay it on the newspaper to dry.

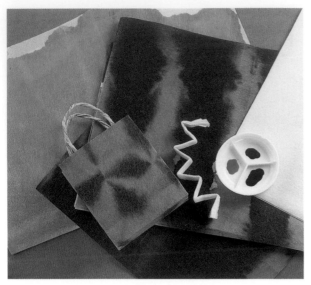

6 Once dry, place the dyed paper between two sheets of clean white paper and iron gently on a low heat: this will set the dye in the paper. Even after ironing the pleat lines will not disappear altogether and should be regarded as part of the pattern.

Different patterns

1 As you dye more pieces of paper experiment with different techniques. For example, try dampening the paper with water before dying it to soften the finished pattern; or let a piece of paper dry, then fold it and dye it again. Tie string or thin wire around the folded paper before dying, so that the pressure points create patterns on the paper.

Making gift bags

1 Glue thin card or good quality paper on to the back of the dyed paper. Make sure that the paper grains run the same way, or the paper may crinkle as the glue dries.

2 Using the measurements for the gift bag on page 56, draw the plan on to the reverse side of the backed paper, making sure that all the angles are right angles; or if you prefer you can use a photocopier to enlarge the plan and use it as a template. Cut out the bag, and score along the dotted lines.

3 Fold over the top flaps and glue them down. Fold up the bag and glue the side seams together. When dry, glue the bottom flaps together. Fold the bag flat, holding it together with a clothes peg until dry.

4 Cut two 30cm (12in) lengths of raffia. Glue the ends inside the bag to form the handles. Glue a piece of paper over the raffia ends to hold them securely inside the bag.

5 For the gift tag, cut a rectangle of backed and dyed paper. Fold in half and punch a hole in the top. Thread a length of raffia through the hole and tie the tag to the bag.

Making the box

1 Glue thin card or paper on to the back of two A4 pieces of dyed paper.

2 Using the measurements for the box base on page 57, draw the plan on to the reverse of the backed paper. Cut out the design and score along the dotted lines.

3 Fold the box in half along the dotted line A: this will make the sides of the box double thickness. Glue the sides together. Fold the box along the dotted lines, then glue the base and sides together on the flaps. Glue a square of dyed paper inside the base of the box.

4 To make the box lid, cut a 21x21cm (8¼x8¼in) square of backed dyed paper. Draw diagonal lines from corner to corner to find the exact centre of the square. Fold in each corner to touch the centre point.

5 Fold the top and bottom edges of the square in to the middle to form a long rectangle, then unfold. Fold the opposite sides of the square into the middle and then unfold.

6 Completely unfolded the square of paper: the paper will have a grid of folded lines that match the diagram on page 56. Cut along the lines coloured red.

7 Fold in the corners again, and then fold the edges upwards to form the box sides. Lift up each triangle from the centre of the box, tuck under the pointed corner ends and then flap down the triangles. The lid will now stay together, although you can add a dab of glue under the point of each triangle.

Making the pyramid box

1 Glue thin card or paper on to the back of dyed paper.

2 Draw the plan for the pyramid box on page 56 on to the reverse side of the dyed and backed paper. Cut out the design and score along the dotted lines. Using a hole punch, make a hole in the top corner of each triangle.

3 Cut a length of red ribbon approximately 40cm (15in) and thread it through the holes at the top of each triangular side. Pull up the ribbon, bringing the sides of the pyramid together. Tie the ribbon in a bow.

Making the box lid

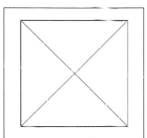

1 Cut a square of paper and draw diagonal lines from corner to corner.

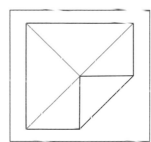

2 Fold in one of the corners of the square to touch the centre point.

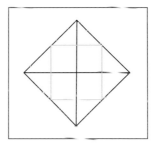

3 Fold in the other three corners to touch the centre point.

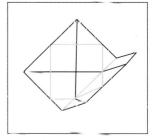

4 Fold the top and bottom edges in, to form a long rectangle, then unfold.

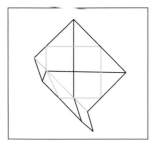

5 Fold in the opposite sides of the square, then unfold.

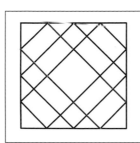

6 Completely unfold the paper, it will now be covered in a grid of lines.

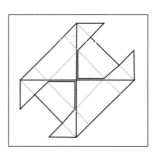

7 Cut along the red lines. Fold the four sides upwards to form the box sides.

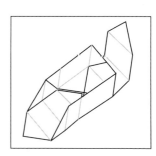

8 Lift up the triangles at the centre point and tuck under the pointed corners.

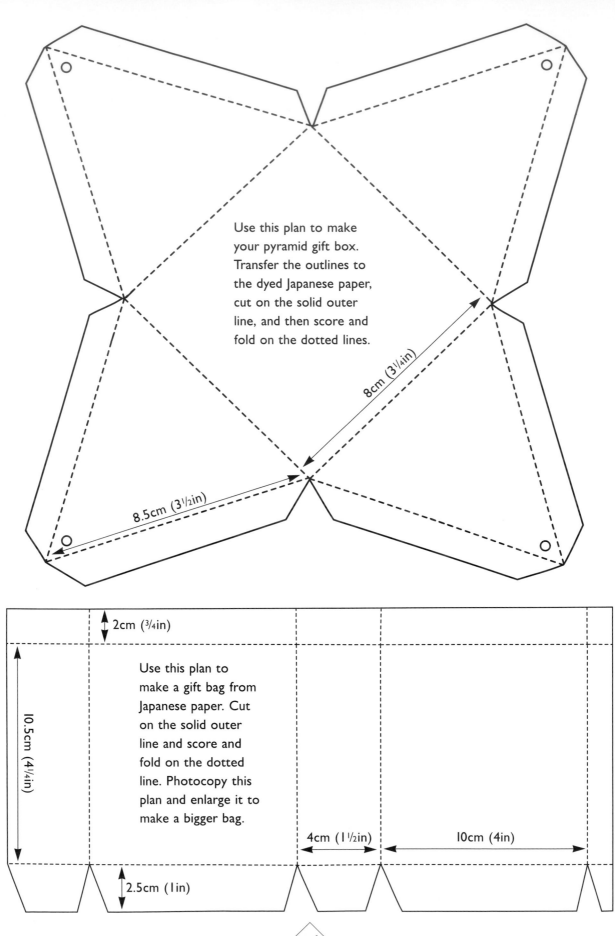

Use this plan to make your pyramid gift box. Transfer the outlines to the dyed Japanese paper, cut on the solid outer line, and then score and fold on the dotted lines.

8cm (3¼in)

8.5cm (3½in)

2cm (¾in)

Use this plan to make a gift bag from Japanese paper. Cut on the solid outer line and score and fold on the dotted line. Photocopy this plan and enlarge it to make a bigger bag.

10.5cm (4¼in)

4cm (1½in)

10cm (4in)

2.5cm (1in)

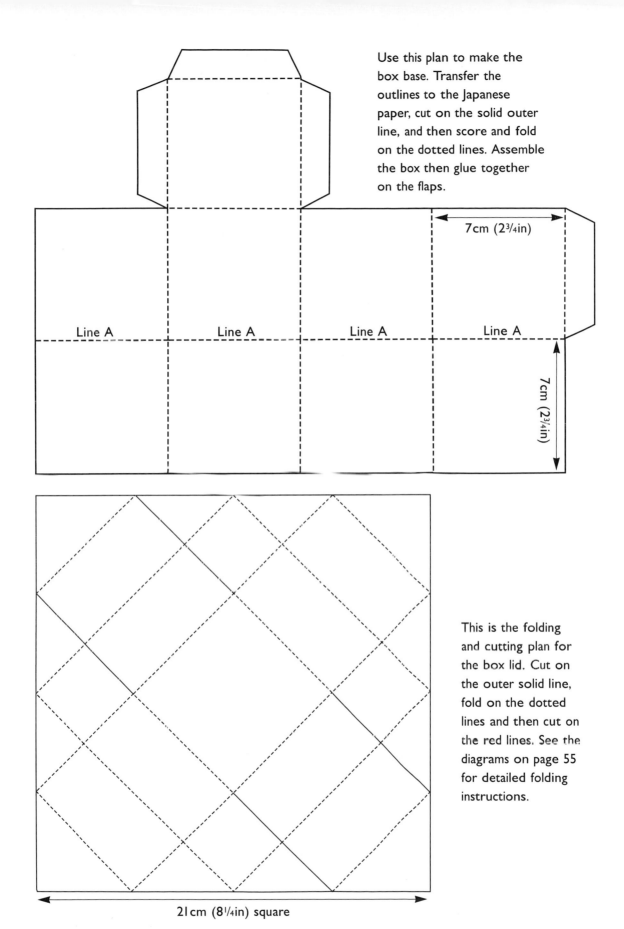

Use this plan to make the box base. Transfer the outlines to the Japanese paper, cut on the solid outer line, and then score and fold on the dotted lines. Assemble the box then glue together on the flaps.

7cm (2¾in)

Line A Line A Line A Line A

7cm (2¾in)

This is the folding and cutting plan for the box lid. Cut on the outer solid line, fold on the dotted lines and then cut on the red lines. See the diagrams on page 55 for detailed folding instructions.

21cm (8¼in) square

Batik Covered Books

Batik is a fun way to decorate plain paper. Patterns are created when melted wax, which has been dropped or brushed on to the paper, is painted over; the wax is then removed, leaving the areas beneath the original paper colour. This process can be repeated several times over, creating some amazing colour schemes and patterns

The instructions for covering the portfolio and book are reversible; the portfolio cover with the triangular corners can be used to cover the book, and the portfolio can be covered with a single sheet of batik paper.

You will need

- Paper – good quality coloured paper or white paper coloured with a colour wash
- Batik wax
- Double boiler or saucepan
- Clean empty pet food tins
- Poster paints
- Book, bookcloth, stiff card
- Ribbon
- Saucers
- Paintbrushes
- Newspaper
- Iron, tacky glue, water
- Ruler, scissors, craft knife, cutting mat

Preparing the paper

1 Either use good quality pale coloured paper, or plain white paper painted with watery poster paint. Allow time for the paper to dry.

Preparing the wax

1 Fill the bottom of a double boiler with water and put batik wax in the top; or put the wax in a clean pet food tin and place the tin in a saucepan of water. Melt the wax on a low heat – never leave the wax unattended, and be careful not to let the water boil dry.

Designing with wax

1 Cover your work surface with newspaper. Squash the sides of a clean pet food tin together to form a spout. Pour some of the melted wax into the tin, and then use it to make patterns on the paper. A crisscross pattern can be made by pouring wax from the tin in parallel lines across the paper, first in one direction and then in the other; pour the wax from the tin in spirals, or let it dribble out to form random spaghetti patterns; make splats by letting the wax fall on to the paper from about 30cm (12in), or use jerking movements to almost throw the wax out of the tin on to the paper. You may prefer a more controlled application by using a paintbrush dipped in melted wax to make streaks and drips across the paper (see Batik Paper, page 15). As the wax touches the paper it will dry almost immediately. You may

need to reheat the wax from time to time to keep it in liquid form.

Colouring the paper

1 Choose a poster paint colour that complements the base colour of your paper. Thin the paint with water and then paint over the batiked paper using long brush strokes. The paint will not stick to the wax pattern; if it does, thin the paint with a little more water. Leave the paper to dry.

Making multicoloured paper

1 Once the wax pattern has been painted over and the paper dried, you can apply another pattern over the top of the first using more wax. Thin a different paint colour with water and then when the wax has set, paint the paper again. You can repeat this process several times, remembering to dry the paper between each layer of paint and wax.

2 Once the paper is dry, sandwich it between several layers of old newspaper. Put the iron on to a cotton setting and iron the paper – the heat from the iron will melt the wax which will be absorbed into the newspaper. This is a very messy process so protect your ironing board and clothing from the melted wax. Keep ironing and changing the newspaper until all the wax has melted.

Making a portfolio

1 Cut two pieces of thick card 22.5x32cm (8³/₄x12¹/₂in) for the covers and one piece 2x32cm (³/₄x12¹/₂in) for the spine. Cut a piece of bookcloth 8x35cm (3¹/₈x13³/₄in) making sure that the warp runs parallel to the spine.

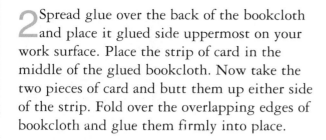

2 Spread glue over the back of the bookcloth and place it glued side uppermost on your work surface. Place the strip of card in the middle of the glued bookcloth. Now take the two pieces of card and butt them up either side of the strip. Fold over the overlapping edges of bookcloth and glue them firmly into place.

3 Cut another piece of bookcloth 8x31.5cm (3¹/₈x12¹/₂in) and glue this on to the inside of the spine. Using the corner of a ruler, rub the edges of the bookcloth making sure it is well stuck down on to the card. Leave the glue to dry.

4 Working on the outside of the portfolio, draw a pencil line 1cm (³/₈in) in from each long edge of the bookcloth. Cut a piece of batik paper 22x35cm (8¹/₈x13¹/₄in) for the front cover; lay the paper on to the cover, lining the long edge up with the pencil line. Cut away a triangle of batik paper at the top and bottom corner of the book, and trim the overlap on the three outer edges to 3mm (¹/₈in). Repeat for the back cover.

5 Glue a triangle of bookcloth on to the four corners of the portfolio, slightly larger than the cut away triangles on the batik paper, and with enough overlap to fold over on to the inside. Glue the paper in position on the covers. Fold the excess paper at the edges over on to the inside, and glue in place.

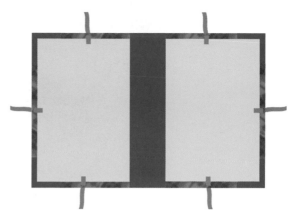

6 To add closing ribbons to the portfolio, you will need to make slots to hold the ribbon. Cut six slots in all – three for each cover. They should be placed in the middle of each side and 1cm (³/₈in) in from the edge. The slots should go right through the paper and the card cover. Cut six pieces of ribbon 25cm (10in) long and thread one through each slot working from the front of the portfolio. Glue the end of each ribbon on to the inside of the cover.

7 To finish the portfolio, glue a piece of plain coloured paper 20x31cm (7³/₄x12¹/₄in) centrally on to the inside of each cover. Place your finished portfolio under a heavy weight, and leave it to dry for a least 24 hours.

Covering a book

1 If possible, try to cover the book in a single sheet of paper. Measure the front and back cover and the spine, adding 3cm (1¹/₄in) to the width and 3cm (1¹/₄in) to the length. Cut a rectangle of batik paper to this size.

2 Open out the book and then glue the batik paper over the spine and the covers, pressing the paper well down to remove any air bubbles. Fold the excess paper at the edges over on to the inside of the book and glue in place.

3 Finish the book in the same way as the portfolio, leaving it to dry for 24 hours.

Acknowledgements

Thanks to the designers for contributing such wonderful projects:
Marbled Daisy Frame (page 16), Jill Millis and Jan Cox
Plant Paper Pictures (page 22), Martin Penny
Paste Decorated Boxes (page 26), Jill Millis
Tinted Paper Lampshades (page 30), Jan Cox and John Underwood
Black Cat Desk Set (page 34), Jill Millis
Glittery Christmas Cards (page 40), Jill Millis
Petal and Leaf Stationery (page 44), Susan Penny
Scrap Paper Photo Albums (page 48), Jan Cox and John Underwood
Fold 'n' Dye Gift Bags (page 52), Jan Cox and John Underwood
Batik Covered Books (page 58), Jan Cox and John Underwood

Many thanks to Design Objectives for supplying the Anita's acrylics, MDF wooden
blanks and Amaco rub and buff (rubbing paste).

Other books in the Made Easy series

Greeting Cards (David & Charles, 2000)

Painted Crafts (David & Charles, 1999)

Candle Making (David & Charles, 1999)

Papier Mâché (David & Charles, 1999)

3-D Découpage (David & Charles, 1999)

Mosaics (David & Charles, 1999)

Ceramic Painting (David & Charles, 1999)

Stamping (David & Charles, 1998)

Stencilling (David & Charles, 1998)

Glass Painting (David & Charles, 1998)

Silk Painting (David & Charles, 1998)

Suppliers

ColArt Fine Art & Graphics Ltd
Whitefriars Avenue
Harrow
Middlesex HA3 5RH
Tel: 0181 4274343
Paint wholesaler, telephone for local stockist
(Acrylic paints)

Design Objectives (Head office only)
36-44 Willis Way
Fleets Industrial Estate
Poole
Dorset BH15 3SU
Tel: 01202 679976
Craft wholesaler, telephone for local stockist
(Acrylic paint, MDF blanks, rubbing paste)

Hobby Crafts (Head office only)
River Court, Southern Sector
Bournemouth International Airport
Christchurch
Dorset BH23 6SE
Tel: 0800 272387 freephone
Retail shops nationwide, telephone for local
store
(Craft warehouse)

Homecrafts Direct
PO Box 38
Leicester LE1 9BU
Tel: 0116 251 3139
Mail order service
(Craft supplies)

Fred Aldous Ltd
37 Lever Street
Manchester
M1 1LW
Tel: 0161 236 3477
Mail order service
(Lampshade frames, fireproof sticky-backed
lining for lampshades)

Seta Silk Paints (Distributor – office only)
Philip and Tacey Ltd
North Way
Andover
Hampshire
SP10 5BA
Tel: 01264 332171
Telephone for your local retail stockist
(Silk paint)

Index